The Force Behind Nature

The Force Behind Nature
and
Man as the Interpreter of Nature

by William B. Carpenter

Part I

The Force Behind Nature

Some years ago, I enjoyed opportunities of discussing with John Stuart Mill many questions of philosophy in which we both felt the deepest interest. Among these was the Doctrine of Causation set forth in his recently published "System of Logic": "We may define the cause of a phenomenon to be the antecedent, or the concurrence of antecedents, on which it is invariably and unconditionally consequent." I pointed out to my friend that, when this assemblage of conditions is analyzed, it is uniformly found resolvable into two categories, which may be distinguished as the *dynamical* and the *material;* the former supplying the *force* or *power* to which the change must be attributed, while the latter affords the *conditions* under which that power is exerted. Thus, I urged, when a man falls from a ladder *because* (as is commonly said) of the breaking of the rung on which his foot was resting, the real or *dynamical* cause of his fall is the force of gravity, or attraction of the earth, which pulls him to the ground when his foot is no longer supported; the loss of support being only the *material condition* or *collocation,* which allowed the force previously acting as *pressure* on the rung to produce the downward *motion* of the man who stood upon it.

To this Mr. Mill's reply was, that the distinction is one of metaphysics, not of logic. I ventured, however, to press on him that, to whichever department of philosophy this point is to be referred, it is one of fundamental importance; that, assuming experience as the basis of our knowledge, we recognize the downward tendency of everybody heavier than air, by our sense of muscular tension in lifting it from the ground, or in resisting its descent toward the earth; and that our cognition of *force* through this form of sensation, being thus quite as immediate and direct as our cognition of *motion* through the visual sense, ought to be equally taken account of.

The promulgation, about the same time, of the doctrine of the "Correlation of the Physical Forces" by Professor (now Sir William) Grove, and the researches of Mr. Joule on the "Mechanical Equivalent of Heat," seemed to me to bring this view of *dynamical* causation into yet greater importance, by showing that what is true of that form of force which produces or resists mechanical (or what is now distinguished as *molar*) motion, may be legitimately extended to those other forms which are manifested in the *molecular* changes that express themselves in chemical action, or impress us with the sensations of heat, light, etc. Partaking of the general ignorance at that time prevalent in this country of the doctrine of "Conservation of Energy," already promulgated in Germany by Mayer and Helmholtz, I myself endeavored to carry Professor Grove's principle into the domain of biology, by showing that what physiologists had been accustomed to call vital force may be regarded as having the same "correlation" with the various forms of physical force as they have with each other.

And in the introduction to the fourth edition of my "Human Physiology" I thus explicitly defined my position:

When this assemblage of antecedents is analyzed, it is uniformly found that they may be resolved into two categories, which may be distinguished as the dynamical and the material, the former supplying the force or power to which the change must be attributed, while the hitter afford the conditions under which that power is exerted. Thus, in a steam-engine we see the dynamical agency of heat made to produce mechanical power by the mode in which it is applied: first, to impart a mutual repulsion to the particles of water; and then, by means of that mutual repulsion, to give motion to the various solid parts of which the machine is composed. And thus, if asked what is the cause of the movement of the steam-engine, we distinguish in our reply between the dynamical condition supplied by the heat and the material condition (or assemblage of conditions) afforded by the "collocation" of the boiler, cylinder, piston, valves, etc. . . . In like manner, if we inquire into the cause of the germination of a seed—which has been brought to the surface of the earth after remaining dormant through having been buried deep beneath the soil for (it may be) thousands of years—we are told that the phenomenon depends upon warmth, moisture, and oxygen; but out of these we single warmth as the dynamical condition, while the oxygen and the water, with the organized structure of the seed itself, and the organic compounds which are stored up in its substance, constitute the material.

The subsequent general recognition by the scientific world of the "correlation" between the forces of nature (under whatever form expressed) has thus given a breadth of foundation to the dynamical doctrine of causation which it previously lacked; and the doctrine, having been afterward formally developed by Professor Bain, was summarized by John Stuart Mill, in the later editions of his "Logic," almost in the very terms in which I had originally propounded it to him in conversation, and had publicly expressed it in the extract just cited: "The chief practical conclusion drawn by Professor Bain, bearing on causation, is that we must distinguish, in the assemblage of conditions which constitutes the cause of a phenomenon, two elements: one, the presence of a force; the other, the collocation or position of objects which is required in order that the force may undergo the particular transmutation which constitutes the phenomenon."

Mr. Mill himself still preferred, however, to express the principle in terms of motion rather than in terms of force: "If the effect, or any part of the effect, to be accounted for consists in putting matter in motion, then any of the objects present which has lost motion has contributed to the effect; and this is the true meaning of the proposition that the cause is that one of the antecedents which exerts active force." As this mode of expressing the facts is sanctioned by high authorities at the present time, it may be well for me to explain more fully the basis of my original contention, that our cognition of force is quite as immediate and direct as our cognition of motion; in fact (as I think I shall be able to prove), even more fundamental, inasmuch as our cognition of matter itself is in great degree dependent upon it.

It has been recently well said that "all true science involves both the knowledge of nature and the knowledge of man; it includes the study of mind, as well as of matter. A philosopher may pursue either, but he can have no complete knowledge of what he investigates, without borrowing from the other department of investigation."

Many of the Nature-philosophers who affirm that we have no knowledge of anything but the matter and motion which lie within the range of "experience" show themselves very imperfectly acquainted with what "experience" really means; unhesitatingly ranking as actual objective facts their own mental interpretations of the sensory impressions they receive from external objects. Many metaphysicians, on the other hand, have reasoned as if our concern were with mental operations alone, and as if the abstractions in which they deal had an existence per se, without any relation to the phenomena of nature. But, among the ablest thinkers of the present time, there seems to be now a pretty general recognition of the necessity for the replacement of the abstract definitions of metaphysics—so far, at least, as they relate to the external world—by psychological expressions of the modes in which the human ego is affected by its changes. Thus the ordinary metaphysical definition of "matter" is that which possesses "extension." But, for this definition to convey any definite idea to our minds, we must know what "extension" means; and this, we are told, is the "occupation of space." Now, the conception of "space," in the opinion of most psychologists, is ordinarily derived from our interpretation of visual sensations; and yet these may be altogether deceptive. When we look at a window from a short distance, we cannot tell by the use of our eyes alone whether the space included by its frame is void, or is occupied by a perfectly transparent and colorless glass. A glass globe is held up in front of it, and we cannot tell by looking at it whether it is empty, or is filled with pure water or some other transparent colorless liquid. And we can take no cognizance by our vision of the atmosphere which surrounds us, unless its transparence is interfered with by mist or fog. Clearly, then, our visual sense cannot per se furnish us with a satisfactory definition of matter.

Now that we have got rid of the fiction of "imponderables," we might fall back on a definition of matter—in use before that fiction was invented—as that which possesses "ponderosity" or weight. But what is weight? The downward tendency, it may be replied, in virtue of which all unsupported bodies fall to the earth. But what is this "tendency"? We might see any number of bodies falling to the ground, and might frame a correct law of their motion, without having the remotest conception of their possessing that downward pressure which we at once recognize when we take a lump of lead or iron into our hands; and it is obviously on our cognition of this pressure, that our idea of weight or ponderosity is based. Now, the instrumentality through which we take cognizance of it seems to me to be threefold: In the first place, we have the sense of simple pressure on the tactile surface—as when, the hand passively resting on a table, a weight is laid upon it. Secondly, we recognize it by the sense of tension which we experience when a weight is attached to a pendent limb, and which we refer to the muscles and ligaments which are thus put on the stretch; or when, the hand resting on the top of a cylinder of glass placed over an air-pump, the air is exhausted from beneath, so as to make us feel the downward "pressure of the atmosphere." In these two cases, the mind is the passive recipient of the sensory impressions. But, thirdly, when we determinately lift a weight or hold it suspended by our hands, we experience, in addition to the sense of pressure and the sense of tension, a sense of effort, which we recognize as an immediate revelation of consciousness, not referable to any physical impression, but of the same kind as that which we experience in a purely mental act, such as the fixation of the attention. And a little consideration will, I think, make it clear that it is on this "sense of effort" in resisting downward pressure that our cognition of weight is essentially based.

For, in the first place, the continuance of a moderate pressure on the cutaneous surface, like other sensory impressions that become habitual, soon ceases to affect us sensorially; for we cognosce rather the changes in the states of our sense-organs than the states themselves. Or, again, we may suffer under a temporary or permanent paralysis of the cutaneous sense, that may prevent our feeling the contact of the body we are lifting or supporting; and yet, recognizing its downward pressure in other ways, we can put our muscles into action to antagonize it. But, secondly, this paralysis may extend to the muscular sense, so that the feeling of muscular tension is wanting as well as that of contact-pressure; and yet none the less can a weight be lifted or sustained by a conscious effort, provided that the deficiency of the guiding sensations ordinarily derived from the muscle itself is supplied by the sight. A woman whose arm is sensorially but not motorially paralyzed can hold up her child as long as she looks at it, and a man affected with the like paralysis of his legs can stand and walk while looking at his feet. But, thirdly, since the mental sense of effort is experienced in every determinate exercise of our muscular power, and is, as all experience teaches, a necessary condition of that exercise; since, again, it is proportioned to the exertion we put forth, and continues as long as that exertion is sustained—it is in this, and not in the cutaneous or muscular impressions which are (so to speak) accidental, that (as it seems to me) we find the real basis of our cognition of the "ponderosity" of matter.

But "ponderosity" cannot be considered an essential property of matter, being merely the "accident" of the earth's attraction for bodies lying within its range. This attraction varies with the distance of a body from the center of the earth; and a body occupying the common center of gravity of the earth and sun would be equally drawn toward both, and would consequently have no "weight." We must, therefore, seek a satisfactory definition of matter elsewhere; and we find the clew to it in the consideration that the sense of effort we experience in antagonizing the downward pressure of a body is but a particular case of our more general cognition of resistance. When we project our hand against a hard and fixed solid body, our consciousness of its resistance to our pressure is exactly that which we experience when we try to raise a weight that we have not strength to lift; while, if that solid be either yielding in its parts or movable as a whole, we measure its resistance, as in lifting a weight, by our sense of the effort necessary to overcome it. When we move our hand through a liquid, we are conscious of a resistance to its motion, which is greater or less according to the "viscosity" of the liquid. And, when we move our open hand through air at rest, we are still conscious of a resistance, our sense of it being augmented by an extension of the surface moved, as in the act of fanning; while, if the air is in motion, we feel its pressure on the sail of a boat by the "pull" of the sheet we hold in our hand, or on the sails of a windmill by the rotation it imparts, the force of which we can estimate by the effort we must put forth to resist it. Attenuate any kind of air or gas as we may, its resistance can still be made apparent by the like communication of its own motion to solid bodies. Thus, in Mr. Crookes's wonderful radiometer, a set of vanes, poised on a pivot within a globe of glass exhausted to a millionth of its ordinary gaseous contents, is whirled round by the movement excited in the molecules of that residual millionth, either by the heat of the radiant beam falling on the surface of the globe or by the passage of an electric current across its interior; and the mechanical force required to impart that motion can be measured with precision, by bringing it into comparison with some other force (as that of gravity) of which we can take immediate cognizance. And thus, as Herbert Spencer remarks, by the decomposition of our knowledge of any form of matter into simpler and simpler components, we must come at last to the simplest, to the ultimate material, to the substratum; and this we find in the impression of resistance we receive through what we may call our "force-sense."

Such being the teachings alike of general and of scientific experience, I can not but feel surprised that any persons claiming the title of philosophers should affirm that we know nothing except matter and motion, and that force is a creation of our own imagination. One might suppose such persons to be either destitute of the "force-sense," or to have based their philosophical system upon the movements of the heavenly bodies which they can only see, instead of upon those mundane phenomena in the cognition of which they can bring their hands to the assistance of their eyes. How essential this assistance is to the formation of correct conceptions of the solid forms and relative positions of the objects around us is known to everyone who has studied the physiology of the senses. Should we not think it absurd on the part of any one who possesses in the use of his hands the means of detecting the error of his visual perceptions, if he were to base a superstructure of reasoning—still more to found a whole system of philosophy—upon the latter alone? Yet such appears to me to be the position of those who deny our direct cognition of force.

Let us suppose (if possible) a man who had enjoyed the full use of his eyes, but whose limbs had been completely paralyzed from infancy, looking on at a game of billiards. He would see a succession of motions connected by regular sequence—the motion of the arm of the player, the stroke of the cue, the roll of the ball, its contact with another ball, the movement of the second ball, the change of direction or the entire stop of the first, the rebound of balls from the cushion in altered directions, and so on.

And he might frame a statement in "terms of motion" of all that passes before his eyes, thinking this all he can know. But suppose the limbs of such a man to be suddenly endowed with the ordinary powers of sensation and movement; let him take the cue into his hands and himself strike the ball; let him hold his hand on the-table so that the rolling ball shall strike it and make him feel its impact; let him hold the second ball and feel the shock imparted to it by the stroke of the first. Can anyone deny that he would thus acquire a dynamical conception linking together the whole succession of phenomena, which he was previously quite incapable of forming; that this dynamical conception is quite as directly based upon the experience derived through his "force-sense" as his kinetic expression was upon that derived through his visual sense; and that this cognition of the force producing the motions is, therefore, fully as much entitled to be introduced into a logical doctrine of causation as the visual cognition of the motions themselves? If it be replied that we have no proof that the movement of the ball we strike is produced by the force which we consciously exert in striking it, I simply rejoin that we have as much proof of it as we have of anything which rests upon universal experience, and which we can verify experimentally as often as we choose to try—quite as much as we have of the existence of anything whatever that is external to ourselves.

Let us take, again, the simple case of magnetic attraction. A man who knows nothing of magnetism sees a piece of iron, brought within a certain distance of what looks looks like a horseshoe bar of the same metal, suddenly jump toward its approximated ends; and might, as before, correctly express the fact in "terms of motion." But let him take the piece of iron in his hands, so as to feel the "pull" upon it when brought sufficiently near the magnet, and he then becomes conscious, through his force-sense, of a power of which he was before utterly ignorant.

Thus, as it seems to me, an analysis of those physical experiences on which all our cognitions of the physical universe around us are really based irresistibly lands us in the conclusion that, as Herbert Spencer expresses it, "All the sensations through which the external world is known to us are explicable by us only as resulting from certain forms of force"; the direct derivation of our conception of force from our own experience of muscular tension (or, as I should myself say, from our own sense of effort) being "a fact which no metaphysical quibbling can set aside." In the words of the able American writer I have already quoted, "The conception of force is one of those universal ideas which belong of necessity to the intellectual furniture of every human mind." By no one has the principle for which I am contending, been more clearly and more authoritatively expressed than by Sir John Herschel, a philosopher who united to his wonderful grasp of Nature-phenomena a profound insight into the action of the mind of man in the interpretation of them:

Whatever attempts have been made by metaphysical writers to reason away the connection of cause and effect, and fritter it down into the unsatisfactory relation of habitual [unconditional] sequence, it is certain that the conception of some more real and intimate connection is quite as strongly impressed upon the human mind as that of the existence of an external world, the vindication of whose reality has, strange to say, been regarded as an achievement of no common merit in the annals of this branch of philosophy. It is our own immediate consciousness of effort, when we exert force to put matter in motion or to oppose and neutralize force, which gives us this internal conviction of power and causation, so far 'as it relates to the material world.—(Treatise on "Astronomy" in Lardner's "Cyclopædia," p. 232.)

Man's position as the "interpreter of Nature" may be not inaptly likened (as it seems to me) to that of an intelligent observer of the working of a cotton-factory, with whose mechanical arrangements he is entirely unacquainted, and of whose moving power he knows nothing whatever. He is taken into a vast apartment, in which he is at first utterly bewildered by the number and variety of the movements going on around him; but, by directing his attention to the several machines, seriatim, he is able to arrive at a classification of them, according to the kind of work which it does. Thus he finds one set carding the cotton-wool supplied to it, so that its confused tangle gives place to a parallel laying of the fibers. He would see another taking up the bundles of carded wool, and drawing them out (after repeated doublings to secure uniformity) into a long, soft cord. This cord he would then trace into the roving-machine, which, by a continuation of the drawing process, further reduces its thickness, at the same time giving it a slight twist to increase its tenacity, so that it admits of being-then wound upon bobbins. Thence he would trace the cord into the spinning-machine, which at the same time stretches and twists the cord, producing from it a yarn whose fineness might vary considerably in different machines. Finally, he would see the spun yarn carried, some as weft and some as woof, into the power-loom, from which it emerges as woven cloth—the final resultant of the whole series of operations.

Concentrating now his attention upon any one of these machines, he studies its wheels, levers, and other moving parts, and tries to comprehend their several actions and the bearing of these upon each other. By long and scrutinizing observation he masters the whole series of sequences, and traces the distribution of motion from a single large axis, through the hundreds (it may be) of separate pieces of the machine directly or indirectly connected with it; and he might thus frame a description of the working of the machine, which might be perfectly correct so far as it goes, and which yet would be defective in one most essential particular—the statement of the force or power by which it is moved. For, so far as mere visual observation could teach him, the machine might be self-moving; and he might thus attribute to each kind an inherent power of carding, roving, drawing, spinning, or weaving, as the case might be.

Carrying his observations further, and noticing that one or another of these machines comes to a standstill, but resumes its motion after an interval, he may include this occasional suspension also in his general expression; but, perplexed by the want of any regularity in its intervals, he will seek some further explanation. Continuing his patient watch, he will see that the stoppage of the machine follows the pulling of a handle by the man in attendance upon it, and that, when the handle is pulled the other way, the machine goes on again; and thus he will be led to introduce a certain position of this handle as one of the antecedent conditions of the machine's action. Still pursuing-his inquiries, he finds out that the axes of the several machines are all in mechanical relation with one great longitudinal shaft, being connected with it either by continuous bands passing around pulleys, or by trains of wheelwork; and at last he discovers the important fact that the movement of the handle which stops the machine breaks the continuity of that relation, shifting a strap from a "fast" to a "loose" pulley, or throwing the wheel work "out of gear"; while the converse movement, which restores that continuity, is followed by the renewed action of the machine, which goes on until the continuity is again broken. Thus he will be led to regard its maintenance as essential to the working of the machine; but nothing that he has yet learned explains to him why it is essential. He has only got at the material collocation which his educated vision enables him to recognize; and, for anything he knows to the contrary, the change in that collocation may be in itself adequate to determine the result.

But let him lay hold of the band which stretches between the main shaft and the axis of one machine, or attempt to stay with his hand the rotation of the train of wheels which connects it with another—he then at once becomes conscious, through his "force-sense," of the power which the band or the wheelwork is the instrument of conveying; and as he finds that the "pull" upon his hand is just the same whether the machine is in motion or not, provided that the band or wheel remains in mechanical connection with the main shaft, he comes to the conviction that the source of the power is in the shaft, and that, so far from any one of the machines having an inherent power of movement, its motion entirely depends upon the force supplied to it from the shaft. And when, under the guidance of this conception, he again examines the working of the several kinds of machine, he finds that, while the power is the same for all, the diversity in their respective products is traceable to the diversity in their construction—that is, to the material collocations through which the one moving force exerts itself in action.

But, having thus acquired the notion of moving power, and having satisfied himself of the derivation of the force that gives motion to each of the entire aggregate of machines, from one main shaft, our inquirer finds himself again posed. Has this shaft itself an inherent power of motion; or does it derive that power from any ulterior source?

He sees the shaft apparently terminate in the two end-walls of the building; and, finding no evidence of its connection with anything else, he may feel himself drawn toward the conclusion that it moves of itself—that is, by the "potency" of its own material constitution. But, before adopting this rationale, he sees all the machines stop at once, and finds that the shaft also has ceased to revolve. Here is a new and startling phenomenon. After pondering on it for an hour, and carefully looking out for an explanation, he sees the shaft and its connected machines resume their motion, and yet is certain that no agency visible to him has had any concern in that renewal. By continued watching, he finds this suspension and renewal to be periodical, so that he can frame a law that shall express them in terms of time. Thus he might give a complete phenomenal account of the action of the shaft which should be perfectly consistent with the assumption of its "inherent potency," and which might be sufficiently satisfactory to his mind to justify him in believing that there is no more to be learned about it. But, not wishing to leave anything uninvestigated, he goes round to the other side of the wall. There he finds that one end of the shaft comes through it, and is in mechanical connection with either a steam-engine or a water-wheel; and, by watching what occurs when its motion is checked and renewed, he sees that the engineer shuts off, or turns on, either the steam generated in the boiler of the steam engine, or the descending water whose motion drives the wheel.

I shall not weary the patience of such readers as may have followed me thus far, by tracing out in like detail the further steps of the inquiry, but shall land them in the final conclusion now accepted by every man of science—that the power exerted in both these cases is drawn from solar radiation: the fall of the water which gives motion to the water-wheel being merely the return of that which has been pumped up as vapor by the sun's heat; while the combustion of coal from which steam-power is derived reproduces, as active force, or "energy," the sunshine that exerted itself during the Carboniferous period in dissociating carbonic acid and water into the hydrocarbons of coal and the oxygen of the atmosphere, whose recombination gives forth heat and light. And, if we look still further back for the source of the sun's radiant energy, we should find it, perhaps, in the progressive consolidation of the primeval "fire-mist"—nebular matter.

But whence nebular matter? And whence the force which draws its particles together, and which manifests itself as light and heat during their consolidation? Here we come to a wall, to the other side of which we seem at present to have no access.

But is there no other side? Does not the whole course of the preceding inquiry show the unsatisfaction (if I may revive an obsolete word) of resting in any inherent "potency" of matter as the ultima ratio of the existing cosmos? If we think the man foolish who supposes the main shaft of a cotton-mill to turn of itself merely because he sees it apparently end in a wall which conceals from him the source of its motive power, are we not really chargeable with the like folly if we attribute self-motion to the ultimate molecules of matter, merely because the power that moves them is hid from our sight? The mere physicist may see no possible way further. But there is a philosophy which has fully as true and as broad a basis in man's psychical experience as can be claimed for the fabric of physical science; and, in the admirable words of the great master I have already quoted, I shall sum up an argument which this paper is intended rather to illustrate and enforce by an appeal to the familiar facts of consciousness than to present in strict logical form:

In the mental sense of effort, clear to the apprehension of everyone who has ever performed a voluntary act, which is present at the instant when the determination to do a thing is carried out into the act of doing it, we have a consciousness of immediate and personal causation which cannot be disputed or ignored. And, when we see the same kind of act performed by another, we never hesitate in assuming for him that consciousness which we recognize in ourselves; and in this case we can verify our conclusion by oral communication. ... In the only case in which we are admitted into any personal knowledge of the origin of force, we find it connected (possibly by intermediate links untraceable by our faculties, yet indisputably connected) with volition, and, by inevitable consequence, with motive, with intellect, and with all those attributes of mind in which personality consists.

As a physiologist, I most fully recognize the fact that the physical force exerted by the body of man is not generated de novo by his will, but is derived from the oxidation of the constituents of his food. But holding it as equally certain, because the fact is capable of verification by everyone as often as he chooses to make the experiment, that, in the performance of every volitional movement, that physical force is put in action, directed, and controlled by the individual personality or ego, I deem it just as absurd and illogical to affirm that there is no place for a God in nature, originating, directing, and controlling its forces by his will, as it would be to assert that there is no place in man's body for his conscious mind.

Part II

Man as the Interpreter of Nature

It has been customary with successive occupants of this chair to open the proceedings of the meetings over which they respectively presided with a discourse on some aspect of Nature in her relation to man. But I am not aware that any one of them has taken up the other side of the inquiry—that which concerns man as the "Interpreter of Nature;" and I have therefore thought it not inappropriate to lead you to the consideration of the mental processes by which are formed those fundamental conceptions of matter and force, of cause and effect, of law and order, which furnish the basis of all scientific reasoning, and constitute the *Philosophia prima* of Bacon. There is a great deal of what I cannot but regard as fallacious and misleading philosophy—"oppositions of science, falsely so called"—abroad in the world at the present time. And I hope to satisfy you that those who set up their own conceptions of the orderly sequence which they discern in the phenomena of Nature as fixed and determinate laws, by which those phenomena not only are, but always have been, and always must be, invariably governed, are really guilty of the intellectual arrogance they condemn in the systems of the ancients, and place themselves in antagonism to those real philosophers by whose grasp and insight that order has been so far disclosed. For what love of the truth, as it is in Nature, was ever more conspicuous than that which Kepler displayed in his abandonment of each of the conceptions of the planetary system which his imagination had successively devised, so soon as it proved to be inconsistent with the facts disclosed by observation? In that almost admiring description of the way in which his enemy Mars, "whom he had left at home a despised captive," had "burst all the chains of the equations, and broke forth from the prisons of the tables," who does not recognize the justice of Schiller's definition of the real philosopher—as one who always loves truth better than his system? And when at last he had gained the full assurance of a success so complete that (as he says) he thought he must be dreaming, or that he had been reasoning in a circle, who does not feel the almost sublimity of the self-abnegation with which, after attaining what was in his own estimation such a glorious reward of his life of toil, he abstains from claiming the applause of his contemporaries, but leaves his fame to after-ages in these noble words: "The book is written; to be read either now or by posterity, I care not which. It may well wait a century for a reader, as God has waited 6,000 years for an observer."

And when a yet greater than Kepler was bringing to its final issue that grandest of all scientific conceptions, long pondered over by his almost superhuman intellect—which linked together the Heavens and the Earth, in the *nexus* of a universal attraction, establishing the truth for whose utterance Galileo had been condemned, and giving to Kepler's Laws a significance of which their author had never dreamed—what was the meaning of that agitation which prevented the philosopher from completing his computation, and compelled him to hand it over to his friend? That it was not the thought of his own greatness, but the glimpse of the grand universal order thus revealed to his mental vision, which shook the soul of Newton to its foundations, we have the proof in that comparison in which he likened himself to a child picking up shells on the shore of the vast ocean of truth—a comparison which will be evidence to all time at once of his true philosophy and of his profound humility.

Though it is with the intellectual representation of Nature which we call science that we are primarily concerned, it will not be without its use to cast a glance in the first instance at the other two principal characters under which man acts as her interpreter—those, namely, of the artist and of the poet.

The artist serves as the interpreter of Nature, not when he works as the mere copyist, delineating that which he sees with his bodily eyes, and which we could see as well for ourselves, but when he endeavors to awaken within us the perception of those beauties and harmonies which his own trained sense has recognized, and thus impart to us the pleasure he has himself derived from their contemplation. As no two artists agree in the original constitution and acquired habits of their minds, all look at Nature with different (mental) eyes; so that, to each, Nature is what he individually sees in her.

The poet, again, serves as the interpreter of Nature, not so much when by skilful word-painting (whether in prose or verse) he calls up before our mental vision the picture of some actual or ideal scene, however beautiful, as when, by rendering into appropriate forms those deeper impressions made by the Nature around him on the moral and emotional part of his own nature, he transfers these impressions to the corresponding part of ours. For it is the attribute of the true poet to penetrate the secret of those mysterious influences which we all unknowingly experience; and, having discovered this to himself, to bring others, by the power he thus wields, into the like sympathetic relation with Nature, evoking with skilful touch the varied response of the Soul's finest chords, heightening its joys, assuaging its griefs, and elevating its aspirations. While, then, the artist aims to picture what he sees in Nature, it is the object of the poet to represent what he feels in Nature; and to each true poet Nature is what he individually finds in her.

The philosopher's interpretation of Nature seems less individual than that of the artist or the poet, because it is based on facts which any one may verify, and is elaborated by reasoning processes of which all admit the validity. He looks at the universe as a vast book lying open before him, of which he has in the first place to learn the characters, then to master the language, and finally to apprehend the ideas which that language conveys. In that book there are many chapters, treating of different subjects; and, as life is too short for any one man to grasp the whole, the scientific interpretation of this book comes to be the work of many intellects, differing not merely in the range but also in the character of their powers. But while there are "diversities of gifts," there is "the same spirit." While each takes his special direction, the general method of study is the same for all. And it is a testimony alike to the truth of that method and to the unity of Nature that there is an ever-increasing tendency toward agreement among those who use it aright—temporary differences of interpretation being removed, sometimes by a more complete mastery of her language, sometimes by a better apprehension of her ideas—and lines of pursuit which had seemed entirely distinct or even widely divergent being found to lead at last to one common goal. And it is this agreement which gives rise to the general belief—in many, to the confident assurance—that the scientific interpretation of Nature represents her not merely as she seems but as she really is.

When, however, we carefully examine the foundation of that assurance, we find reason to distrust its security; for it can be shown to be no less true of the scientific conception of Nature than it is of the artistic or the poetic, that it is a representation framed by the mind itself out of the materials supplied by the impressions which external objects make upon the senses, so that, to each man of science, Nature is what he individually believes her to be. And that belief will rest on very different bases, and will have very unequal values, in different departments of science. Thus in what are commonly known as the "exact" sciences, of which astronomy may be taken as the type, the data afforded by precise methods of observation can be made the basis of reasoning, in every step of which the mathematician feels the fullest assurance of certainty; and the final deduction is justified either by its conformity to known or ascertainable facts—as when Kepler determined the elliptic orbit of Mars; or by the fulfillment of the predictions it has sanctioned—as in the occurrence of an eclipse or an occultation at the precise moment specified many years previously; or, still more emphatically, by the actual discovery of phenomena till then unrecognized—as when the perturbations of the planets, shown by Newton to be the necessary results of their mutual attraction, were proved by observation to have a real existence; or, as when the unknown disturber of Uranus was found in the place assigned to him by the computations of Adams and Le Verrier.

We are accustomed, and I think most rightly, to speak of these achievements as triumphs of the human intellect. But the very phrase implies that the work is done by mental agency, and the coincidence of its results with the facts of observation is far from proving the intellectual process to have been correct. For we learn, from the confessions of Kepler, that he was led to the discovery of the elliptic orbit of Mars by a series of happy accidents, which turned his erroneous guesses into the right direction; and to that of the passage of the radius vector over equal areas in equal times by the motion of a whirling force emanating from the sun, which we now regard as an entirely wrong conception of the cause of orbital revolution. It should always be remembered, moreover, that the Ptolemaic system of astronomy, with all its cumbrous ideal mechanism, did intellectually represent all that the astronomer, prior to the invention of the telescope, could see from his stand-point the earth with an accuracy which was proved by the fulfillment of his anticipations. And in that last and most memorable prediction, which has given an imperishable fame to our two illustrious contemporaries, the inadequacy of the basis afforded by actual observation of the perturbations of Uranus required that it should be supplemented by an assumption of the probable distance of the disturbing planet beyond, which has been shown by subsequent observation to have been only an approximation to the truth.

Even in this most exact of sciences, therefore, we cannot proceed a step without translating the actual phenomena of Nature into intellectual representations of those phenomena, and it is because the Newtonian conception is not only the most simple, but is also, up to the extent of our present knowledge, universal in its conformity to the facts of observation, that we accept it as the only scheme of the universe yet promulgated which satisfies our intellectual requirements.

When, under the reign of the Ptolemaic system, any new inequality was discovered in the motion of a planet, a new wheel had to be added to the ideal mechanism—as Ptolemy said, "to save appearances." If it should prove, a century hence, that the motion of Neptune himself is disturbed by some other attraction than that exerted by the interior planets, we should confidently expect that not an ideal but a real cause for that disturbance will be found in the existence of another planet beyond. But I trust that I have now made it evident to you that this confident expectation is not justified by any absolute necessity of Nature, but arises entirely out of our belief in her uniformity; and into the grounds of this and other primary beliefs, which serve as the foundation of all scientific reasoning, we shall presently inquire.

There is another class of cases, in which an equal certainty is generally claimed for conclusions that seem to flow immediately from observed facts, though really evolved by intellectual processes; the apparent simplicity and directness of those processes either causing them to be overlooked or veiling the assumptions on which they are based. Thus Mr. Lockyer speaks as confidently of the sun's chromosphere of incandescent hydrogen, and of the local outbursts which cause it to send forth projections tens of thousands of miles high, as if he had been able to capture a flask of this gas, and had generated water by causing it to unite with oxygen.

Yet this confidence is entirely based on the assumption that a certain line which is seen in the spectrum of a hydrogen-flame means hydrogen also when seen in the spectrum of the sun's chromosphere; and, high as is the probability of that assumption, it cannot be regarded as a demonstrated certainty, since it is by no means inconceivable that the same line might be produced by some other substance at present unknown. And so, when Dr. Huggins deduces, from the different relative positions of certain lines in the spectra of different stars, that these stars are moving from or toward us in space, his train of reasoning is based on the assumption that these lines have the same meaning—that is, that they represent the same elements—in every luminary. That assumption, like the preceding, may be regarded as possessing a sufficiently high probability to justify the reasoning based upon it; more especially since, by the other researches of that excellent observer, the same chemical elements have been detected as vapors in those filmy cloudlets which seem to be stars in an early stage of consolidation. But, when

Frankland and Lockyer, seeing in the spectrum of the yellow solar prominences a certain bright line not identifiable with that of any known terrestrial flame, attribute this to an hypothetical new substance which they propose to call Helium, it is obvious that their assumption rests on a far less secure foundation, until it shall have received that verification which, in the case of Mr. Crooke's researches on Thallium, was afforded by the actual discovery of the new metal, the presence of which had been indicated to him by a line in the spectrum not attributable to any substance then known.

In a large number of other cases, moreover, our scientific interpretations are clearly matters of judgment; and this is eminently a personal act, the value of its results depending in each case upon the qualifications of the individual for arriving at a correct decision. The surest of such judgments are those dictated by what we term "common-sense," as to matters on which there seems no room for difference of opinion, because every sane person comes to the same conclusion, although he may he able to give no other reason for it than that it appears to him "self-evident." Thus, while philosophers have raised a thick cloud of dust in the discussion of the basis of our belief in the existence of a world external to ourselves—of the Non-Ego, as distinct from the Ego—and while every logician claims to have found some flaw in the proof advanced by every other, the common-sense of mankind has arrived at a decision that is practically worth all the arguments of all the philosophers who have fought again and again over this battle-ground. And I think it can be shown that the trustworthiness of this common-sense decision arises from its dependence, not on any one set of experiences, but upon our unconscious coordination of the whole aggregate of our experiences—not on the conclusiveness of any one train of reasoning, but on the convergence of all our lines of thought toward this one centre.

Now, this "common-sense," disciplined and enlarged by appropriate culture, becomes one of our most valuable instruments of scientific inquiry; affording in many instances the best, and sometimes the only, basis for a rational conclusion. Let us take as a typical case, in which no special knowledge is required, what we are accustomed to call the "flint implements" of the Abbeville and Amiens gravel-beds. No logical proof can be adduced that the peculiar shapes of these flints were given to them by human hands; but does any unprejudiced person now doubt it? The evidence of design, to which, after an examination of one or two such specimens, we should only be justified in attaching a probable value, derives an irresistible cogency from accumulation. On the other hand, the improbability that these flints acquired their peculiar shape by accident becomes to our minds greater and greater as more and more such specimens are found; until at last this hypothesis, although it cannot be directly disproved, is felt to be almost inconceivable, except by minds previously "possessed" by the "dominant idea" of the modern origin of man. And thus what was in the first instance a matter of discussion has now become one of those "self-evident" propositions which claim the unhesitating assent of all whose opinion on the subject is entitled to the least weight.

We proceed upward, however, from such questions as the common-sense of mankind generally is competent to decide, to those in which special knowledge is required to give value to the judgment; and thus the interpretation of Nature by the use of that faculty comes to be more and more individual; things being perfectly "self-evident" to men of special culture which ordinary men, or men whose training has lain in a different direction, do not apprehend as such. Of all departments of science, geology seems to me to be the one that most depends on this specially-trained "common-sense;"

which brings as it were into one focus the light afforded by a great variety of studies—physical and chemical, geographical—and biological and throws it on the pages of that Great Stone Book on which the past history of our globe is recorded. And while Astronomy is of all sciences that which may be considered as most nearly representing Nature as she really is, Geology is that which most completely represents her as seen through the medium of the interpreting mind; the meaning of the phenomena that constitute its data being in almost every instance open to question, and the judgments passed upon the same facts being often different according to the qualifications of the several judges. No one who has even a general acquaintance with the history of this department of science can fail to see that the geology of each epoch has been the reflection of the minds by which its study was then directed; and that its true progress dates from the time when that "common-sense" method of interpretation came to be generally adopted which consists in seeking the explanation of past changes in the forces at present in operation, instead of invoking the aid of extraordinary and mysterious agencies, as the older geologists were wont to do whenever they wanted—like the Ptolemaic astronomers—"to save appearances." The whole tendency of the ever-widening range of modern geological inquiry has been to show how little reliance can be placed upon the so-called "laws" of stratigraphical and paleontological succession, and how much allowance has to be made for local conditions. So that, while the astronomer is constantly enabled to point to the fulfilment of his predictions as an evidence of the correctness of his method, the geologist is almost entirely destitute of any such means of verification. For the value of any prediction that he may hazard—as in regard to the existence or non-existence of coal in any given area—depends not only upon the truth of the general doctrines of geology in regard to the succession of stratified deposits, but still more upon the detailed knowledge which he may have acquired of the distribution of those deposits in the particular locality. Hence no reasonably-judging man would discredit either the general doctrines or the methods of geology, because the prediction proves untrue in such a case as that now about to be brought in this neighborhood to the trial of experience.

We have thus considered man's function as the scientific interpreter of Nature in two departments of natural knowledge, one of which affords an example of the strictest and the other of the freest method which man can employ in constructing his intellectual representation of the universe. And, as it would be found that in the study of all other departments the same methods are used either separately or in combination, we may pass at once to the other side of our inquiry—namely, the origin of those primary beliefs which constitute the groundwork of all scientific reasoning.

The whole fabric of geometry rests upon certain axioms which everyone accepts as true, but of which it is necessary that the truth should be assumed, because they are incapable of demonstration. So, too, the deliverances of our common-sense derive their trustworthiness from what we consider the "self-evidence" of the propositions affirmed. This inquiry brings us face to face with one of the great philosophical problems of our day, which has been discussed by logicians and metaphysicians of the very highest ability as leaders of opposing schools, with the one result of showing how much can be said on each side.

By the intuitionalists it is asserted that the tendency to form these primary beliefs is inborn in man, an original part of his mental organization; so that they grow up spontaneously in his mind as its faculties are gradually unfolded and developed, requiring no other experience for their geneses than that which suffices to call these faculties into exercise. But, by the advocates of the doctrine which regards experience as the basis of all our knowledge, it is maintained that the primary beliefs of each individual are nothing else than generalizations which he forms of such experiences as he has either himself acquired or has consciously learned from others, and they deny that there is any original or intuitive tendency to the formation of such beliefs, beyond that which consists in the power of retaining and generalizing experiences.

I have not introduced this subject with any idea of placing before you even a summary of the ingenious arguments by which these opposing doctrines have been respectively supported; nor should I have touched on the question at all, if I did not believe that a means of reconcilement between them can be found in the idea that the intellectual intuitions of any one generation are the embodied experiences of the previous race. For, as it appears to me, there has been a progressive improvement in the thinking power of man; every product of the culture which has preceded serving to prepare the soil for yet more abundant harvests in the future.

Now, as there can be no doubt of the hereditary transmission in man of acquired constitutional peculiarities, which manifest themselves alike in tendencies to bodily and to mental disease, so it seems equally certain that acquired mental habitudes often impress themselves on his organization, with sufficient force and permanence to occasion their transmission to the offspring as tendencies to similar modes of thought. And thus, while all admit that knowledge cannot thus descend from one generation to another, an increased aptitude for the acquirement, either of knowledge generally or of some particular kind of it, may be thus inherited. These tendencies and aptitudes will acquire additional strength, expansion, and permanence, in each new generation, from their habitual exercise upon the materials supplied by a continually-enlarged experience; and thus the acquired habitudes produced by the intellectual culture of ages will become "a second nature" to everyone who inherits them.

We have an illustration of this progress in the fact of continual occurrence, that conceptions which prove inadmissible to the minds of one generation in consequence either of their want of intellectual power to apprehend them or of their preoccupation by other habits of thought, subsequently find a universal acceptance, and even come to be approved as "self-evident." Thus the first law of motion, divined by the genius of Newton, though opposed by many philosophers of his time as contrary to all experience, is now accepted by common consent, not merely as a legitimate inference from experiment, but as the expression of a necessary and universal truth; and the same axiomatic value is extended to the still more general doctrine that energy of any kind, whether manifested in the "molar" motion of masses, or consisting in the "molecular" motion of atoms, must continue under some form or other without abatement or decay; what all admit in regard to the indestructibility of matter being accepted as no less true of force—namely, that as *ex nihilo nil fit, so nil fit ad nihilum*. But, it may be urged, the very conception of these and similar great truths is in itself a typical example of intuition. The men who divined and enunciated them stand out above their fellows, as possessed of a genius which could not only combine but create, of an insight which could clearly discern what reason could but dimly shadow forth. Granting this freely, I think it may be shown that the intuitions of individual genius are but specially-exalted forms of endowments which are the general property of the race at the time, and which have come to be so in virtue of its whole previous culture. Who, for example, could refuse to the marvelous aptitude for perceiving the relations of numbers, which displayed itself in the untutored boyhood of George Bidder and Zerah Colburn, the title of an intuitive gift? But who, on the other hand, can believe that a Bidder or a Colburn could suddenly arise in a race of savages who cannot count beyond five? Or, again, in the history of the very earliest years of Mozart, who can fail to recognize the dawn of that glorious genius, whose brilliant but brief career left its imperishable impress on the art it enriched? But who would be bold enough to affirm that an infant Mozart could be born among a tribe whose only musical instrument is a tom-tom, whose only song is a monotonous chant?

Again, by tracing the gradual genesis of some of those ideas which we now accept as "self-evident"—such, for example, as that of the "Uniformity of Nature"—we are able to recognize them as the expressions of certain intellectual tendencies, which have progressively augmented in force in successive generations, and now manifest themselves as mental instincts that penetrate and direct our ordinary course of thought. Such instincts constitute a precious heritage, which has been transmitted to us with ever-increasing value through the long succession of preceding generations; and which it is for us to transmit to those who shall come after us, with all that further increase which our higher culture and wider range of knowledge can impart.

And now, having studied the working action of the human intellect in the scientific interpretation of Nature, we shall examine the general character of its products; and the first of these with which we shall deal is our conception of matter and of its relation to force.

The psychologist of the present day views matter entirely through the light of his own consciousness: his idea of matter in the abstract being that it is a "something" which has a permanent power of exciting sensations, his idea of any "property" of matter being the mental representation of some kind of sensory impression he has received from it; and his idea of any particular kind of matter being the representation of the whole aggregate of the sense-perceptions which its presence has called up in his mind. Thus, when I press my hand against this table, I recognize its unyieldingness through the conjoint medium of my sense of touch, my muscular sense, and my mental sense of effort, to which it will be convenient to give the general designation of the tactile sense; and I attribute to that table a hardness which resists the effort I make to press my hand into its substance, while I also recognize the fact that the force I have employed is not sufficient to move its mass. But I press my hand against a lump of dough, and, finding that its substance yields under my pressure, I call it soft. Or, again, I press my hand against this desk, and I find that, although I do not thereby change its form, I change its place; and so I get the tactile idea of motion. Again, by the impressions received through the same sensorial apparatus, when I lift this book in my hand, I am led to attach to it the notion of weight or ponderosity; and, by lifting different solids of about the same size, I am enabled, by the different degrees of exertion I find myself obliged to make in order to sustain them, to distinguish some of them as light, and others as heavy. Through the medium of another set of sense-perceptions, which some regard as belonging to a different category, we distinguish between bodies that feel "hot" and those that feel "cold;" and in this manner we arrive at the notion of differences of temperature. And it is through the medium of our tactile sense, without any aid from vision, that we first gain the idea of solid form, or the three dimensions of space.

Again, by the extension of our tactile experiences, we acquire the notion of liquids, as forms of matter yielding readily to pressure, but possessing a sensible weight which may equal that of solids; and of air, whose resisting power is much slighter, and whose weight is so small that it can only be made sensible by artificial means. Thus, then, we arrive at the notions of resistance and of weight as properties common to all forms of matter; and, now that we have got rid of that idea of light and heat, electricity and magnetism, as "imponderable fluids," which used to vex our souls in our scientific childhood, and of which the popular term "electric fluid" is a "survival," we accept these properties as affording the practical distinction between the "material" and the "immaterial."

Turning, now, to that other great portal of sensation, the sight, through which we receive most of the messages sent to us from the universe around, we recognize the same truth. Thus it is agreed, alike by physicists and physiologists, that color does not exist as such in the object itself; which has merely the power of reflecting or transmitting a certain number of millions of undulations in a second; and these only produce that affection of our consciousness, which we call color, when they fall upon the retina of the living percipient. And if there be that defect, either in the retina or in the apparatus behind it, which we call "color-blindness" or Daltonism, some particular hues cannot be distinguished, or there may even be no power of distinguishing any color whatever. If we were all like Dalton, we should see no difference, except in form, between ripe cherries hanging on a tree and the green leaves around them; if we were all affected with the severest form of color-blindness, the fair face of Nature would be seen by us as in the *chiaro-scuro* of an engraving of one of

Turner's landscapes, not as in the glowing hues of the wondrous picture itself. And, in regard to our visual conceptions, it may be stated with perfect certainty, as the result of very numerous observations made upon persons who have acquired sight for the first time, that these do not serve for the recognition even of those objects with which the individual had become most familiar through the touch until the two sets of sense-perceptions have been coordinated by experience. When once this coordination has been effected, however, the composite perception of form, which we derive from the visual sense alone, is so complete that we seldom require to fall back upon the touch for any further information respecting that quality of the object. So, again, while it is from the coordination of the two dissimilar pictures formed by any solid or projecting object upon our two retinæ that (as Sir Charles Wheatstone's admirable investigations have shown) we ordinarily derive through the sight alone a correct notion of its solid form, there is adequate evidence that this notion also is a mental judgment based on the experience we have acquired in early infancy by the consentaneous exercise of the visual and tactile senses.

Take, again, the case of those wonderful instruments by which our visual range is extended almost into the infinity of space or into the infinity of minuteness. It is the mental, not the bodily, eye that takes cognizance of what the telescope and the microscope reveal to us. For, we should have no well-grounded confidence in their revelations as to the unknown, if we had not first acquired experience in distinguishing the true from the false by applying them to known objects; and every interpretation of what we see through their instrumentality is a mental judgment as to the probable form, size, and movement of bodies removed by either their distance or their minuteness from being cognosced by our sense of touch.

The case is still stronger in regard to that last addition to our scientific *armamentum* which promises to be not inferior in value either to the telescope or the microscope; for it may be truly said of the spectroscope that it has not merely extended the range of our vision, but has almost given us a new sense by enabling us to recognize distinctive properties in the chemical elements which were previously quite unknown. And who shall now say that we know all that is to be known as to any form of matter, or that the science of the fourth quarter of this century may not furnish us with as great an enlargement of our knowledge of its properties, and of our power of recognizing them, as that of its third has done?

But, it may be said, Is not this view of the material universe open to the imputation that it is "evolved out of the depths of our own consciousness"—a projection of our own intellect into what surrounds us—an ideal rather than a real world? If all we know of matter be an "intellectual conception," how are we to distinguish this from such as we form in our dreams, for these, as our Laureate no less happily than philosophically expresses it, are "true while they last." Here our "common-sense" comes to the rescue.

We "awake, and behold it was a dream." Every healthy mind is conscious of the difference between his waking and his dreaming experiences, or, if he is now and then puzzled to answer the question, "Did this really happen or did I dream it?" the perplexity arises from the consciousness that it might have happened. And every healthy mind, finding its own experiences of its waking state not only self-consistent, but consistent with the experiences of others, accepts them as the basis of his beliefs, in preference to even the most vivid recollections of his dreams.

The lunatic pauper, who regards himself as a king, the asylum in which he is confined as a palace of regal splendor, and his keepers as obsequious attendants, is so "possessed" by the conception framed by his disordered intellect that he does project it out of himself into his surroundings; his refusal to admit the corrective teaching of common-sense being the very essence of his malady. And there are not a few persons abroad in the world who equally resist the teachings of educated common-sense whenever they run counter to their own preconceptions, and who may be regarded as—in so far—affected with what I once heard Mr. Carlyle pithily characterize as a "diluted insanity."

It has been asserted over and over again of late years, by a class of men who claim to be the only true interpreters of Nature, that we know nothing but matter and the laws of matter, and that force is a mere fiction of the imagination. May it not be affirmed, on the other hand, that, while our notion of matter is a conception of the intellect, force is that of which we have the most direct—perhaps even the only direct—cognizance? As I have already shown you, the knowledge of resistance and of weight which we gain through our tactile sense is derived from our own perception of exertion; and in vision, as in hearing, it is the force with which the undulations strike the sensitive surface that affects our consciousness with sights or sounds. True it is that in our visual and auditory sensations we do not, as in our tactile, directly cognosce the force which produces them; but the Physicist has no difficulty in making sensible to us, indirectly, the undulations by which sound is propagated, and in proving to our intellect that the force concerned in the transmission of light is really enormous.

It seems strange that those who make the loudest appeal to experience as the basis of all knowledge, should thus disregard the most constant, the most fundamental, the most direct of all experiences; as to which the common-sense of mankind affords a guiding light much clearer than any that can be seen through the dust of philosophical discussion. For, as Sir John Herschel most truly remarked, the universal consciousness of mankind is as much in accord in regard to the existence of a real and intimate connection between cause and effect as it is in regard to the existence of an external world; and that consciousness arises to every one out of his own sense of personal exertion in the origination of changes by his individual agency.

Now, while fully accepting the logical definition of cause as the "antecedent or concurrence of antecedents on which the effect is invariably and unconditionally consequent," we can always single out one dynamical—antecedent the power which does the work—from the aggregate of material conditions under which that power may be distributed and applied. No doubt the term "cause" is very loosely employed in popular phraseology; often (as Mr. Mill has shown) to designate the occurrence that immediately preceded the effect—as when it is said that the spark which falls into a barrel of gunpowder is the cause of its explosion, or that the slipping of a man's foot off the rung of a ladder is the cause of his fall. But even a very slightly-trained intelligence can distinguish the power which acts in each case from the conditions under which it acts. The force which produces the explosion is locked up, as it were, in the powder; and ignition merely liberates it by bringing about new chemical combinations. The fall of the man from the ladder is due to the gravity which was equally pulling him down while he rested on it; and the loss of support, either by the slipping of his foot or by the breaking of the rung, is merely that change in the material conditions which gives the power a new action.

Many of you have doubtless viewed with admiring interest that truly wonderful work of human design, the Walter printing-machine. You first examine it at rest; presently comes a man who simply pulls a handle toward him, and the whole inert mechanism becomes instinct with life—the blank paper, continuously rolling off the cylinder at one end, being delivered at the other, without any intermediate human agency, as large sheets of print, at the rate of 15,000 in an hour. Now, what is the cause of this most marvelous effect? Surely it lies essentially in the power or force which the pulling of the handle brings to bear on the machine from some extraneous source of power, which we in this instance know to be a steam-engine on the other side of the wall. This force it is, which, distributed through the various parts of the mechanism, really performs the action of which each is the instrument; they only supply the vehicle for its transmission and application. The man comes again, pushes the handle in the opposite direction, detaches the machine from the steam-engine, and the whole comes to a stand; and so it remains, like an inanimate corpse, until recalled to activity by the renewal of its moving power.

But, say the reasoners who deny that force is anything else than a fiction of the imagination, the revolving shaft of the steam-engine is "matter in motion;" and, when the connection is established between that shaft and the one that drives the machine, the motion is communicated from the former to the latter, and thence distributed to the several parts of the mechanism. This account of the operation is just what an observer might give who had looked on with entire ignorance of every thing but what his eyes could see; the moment he puts his hand upon any part of the machinery and tries to stop its motion, he takes as direct cognizance, through his sense of the effort required to resist it, of the force which produces that motion as he does through his eye of the motion itself.

Now, since it is universally admitted that our notion of the external world would be not only incomplete, but erroneous, if our visual perceptions were not supplemented by our tactile, so, as it seems to me, our interpretation of the phenomena of the Universe must be very inadequate if we do not mentally coordinate the idea of force with that of motion, and recognize it as the "efficient cause" of those phenomena—the "material conditions" constituting (to use the old scholastic term) only "their formal cause." And I lay the greater stress on this point, because the mechanical philosophy of the present day tends more and more to express itself in terms of motion rather than in terms of force—to become kinetics, instead of dynamics.

Thus, from whatever side we look at this question—whether the common-sense of mankind, the logical analysis of the relation between cause and effect, or the study of the working of our own intellects in the interpretation of Nature—we seem led to the same conclusion—that the notion of force is one of those elementary forms of thought with which we can no more dispense than we can with the notion of space or of succession. And I shall now, in the last place, endeavor to show you that it is the substitution of the dynamical for the mere phenomenal idea which gives their highest value to our conceptions of that order of Nature, which is worshipped as itself a god by the class of interpreters whose doctrine I call in question.

The most illustrative, as well as the most illustrious example of the difference between the mere generalization of phenomena and the dynamical conception that applies to them, is furnished by the contrast between the so-called laws of planetary motion discovered by the persevering ingenuity of Kepler, and the interpretation of that motion given us by the profound insight of Newton. Kepler's three laws were nothing more than comprehensive statements of certain groups of phenomena determined by observation. The first, that of the revolution of the planets in elliptical orbits, was based on the study of the observed places of Mars alone; it might or might not be true of the other planets; for, so far as Kepler knew, there was no reason why the orbits of some of them might not be the eccentric circles which he had first supposed that of Mars to be. So Kepler's second law of the passage of the radius vector over equal areas in equal times, so long as it was simply a generalization of facts in the case of that one planet, carried with it no reason for its applicability to other cases, except that which it might derive from his erroneous conception of a whirling force. And his third law was in like manner simply an expression of a certain harmonic relation which he had discovered between the times and the distances of the planets, having no more rational value than any other of his numerous hypotheses.

Now, the Newtonian "laws" are often spoken of as if they were merely higher generalizations in which Kepler's are included; to me they seem to possess an altogether different character. For, starting with the conception of two forces, one of them tending to produce continuous uniform motion in a straight line, the other tending to produce a uniformly accelerated motion toward a fixed point, Newton's wonderful mastery of geometrical reasoning enabled him to show that, if these dynamical assumptions be granted, Kepler's phenomenal "laws," being a necessary consequence of them, must be universally true. And while that demonstration would have been alone sufficient to give him an imperishable renown, it was his still greater glory to divine that the fall of the moon toward the earth—that is, the deflection of her path from a tangential line to an ellipse—is a phenomenon of the same order as the fall of a stone to the ground; and thus to show the applicability to the entire universe of those simple dynamical conceptions which constitute the basis of the geometry of the "Principia."

Thus, then, while no "law" which is simply a generalization of phenomena can be considered as having any coercive action, we may assign that value to laws which express the universal conditions of the action of a force, the existence of which we learn from the testimony of our own consciousness. The assurance we feel, that the attraction of gravitation must act under all circumstances according to its one simple law, is of a very different order from that which we have in regard (for example) to the laws of chemical attraction, which are as yet only generalizations of phenomena. And yet, even in that strong assurance, we are required, by our examination of the basis on which it rests, to admit a reserve of the possibility of something different—a reserve which we may well believe that Newton himself must have entertained.

A most valuable lesson as to the allowance we ought to make for the unknown "possibilities of Nature" is taught us by an exceptional phenomenon so familiar that it does not attract the notice it has a right to claim. Next to the law of the universal attraction of masses of matter, there is none that has a wider range than that of the expansion of bodies by heat. Excluding water and one or two other substances, the fact of such expansion might be said to be invariable, and, as regards bodies whose gaseous condition is known, the law of expansion can be stated in a form no less simple and definite than the law of gravitation. Supposing those exceptions then to be unknown, the law would be universal in its range. But it comes to be discovered that water, while conforming to it in its expansion from 39 upward to its boiling-point, as also, when it passes into steam, to the special law of expansion of vapors, is exceptional in its expansion also from 39½° downward to its freezing-point; and of this failure in the universality of the law no *rationale* can be given. Still more strange is it that by dissolving a little salt in water we should remove this exceptional peculiarity, for sea-water continues to contract from 39 downward to its freezing-point 12° or 14° lower, just as it does with reduction of temperature at higher ranges.

Thus, from our study of the mode in which we arrive at those conceptions of the orderly sequence observable in the phenomena of Nature which we call "laws," we are led to the conclusion that they are human conceptions subject to human fallibility, and that they may or may not express the ideas of the Great Author of Nature. To set up these laws as self-acting, and as either excluding or rendering unnecessary the power which alone can give them effect, appears to me as arrogant as it is unphilosophical. To speak of any law as "regulating" or "governing" phenomena is only permissible on the assumption that the law is the expression of the *modus operandi* of a governing power. I was once in a great city which for two days was in the hands of a lawless mob. Magisterial authority was suspended by timidity and doubt; the force at its command was paralyzed by want of resolute direction. The "laws" were on the statute-book, but there was no power to enforce them. And so the powers of evil did their terrible work, and fire and rapine continued to destroy life and property without check, until new power came in, when the reign of law was restored.

And thus we are led to the culminating point of man's intellectual interpretation of Nature—his recognition of the unity of the power of which her phenomena are the diversified manifestations. Toward this point all scientific inquiry now tends. The convertibility of the physical forces, the correlation of these with the vital, and the intimacy of that *nexus* between mental and bodily activity which, explain it as we may, cannot be denied, all lead upward toward one and the same conclusion; and the pyramid of which the philosophical conclusion is the apex has its foundation in the primitive instincts of humanity.

By our own remote progenitors, as by the untutored savage of the present day, every change in which human agency was not apparent was referred to a particular animating intelligence. And thus they attributed not only the movements of the heavenly bodies, but all the phenomena of Nature, each to its own deity. These deities were invested with more than human power; but they were also supposed capable of human passions and subject to human capriciousness. As the uniformities of Nature came to be more distinctly recognized, some of these deities were invested with a dominant control, while others were supposed to be their subordinate ministers. A serene majesty was attributed to the greater gods who sit above the clouds; while their inferiors might "come down to earth in the likeness of men." With the growth of the scientific study of Nature, the conception of its harmony and unity gained ever-increasing strength. And so, among the most enlightened of the Greek and Roman philosophers, we find a distinct recognition of the idea of the unity of the directing mind from which the order of Nature proceeds; for they obviously believed that, as our modern poet has expressed it—

All are but parts of one stupendous whole,

Whose body Nature is, and God the Soul."

The science of modern times, however, has taken a more special direction. Fixing its attention exclusively on the order of Nature, it has separated itself wholly from theology, whose function it is to seek after its cause. In this, science is fully justified, alike by the entire independence of its objects, and by the historical fact that it has been continually hampered and impeded, in its search for the truth as it is in Nature, by the restraints which theologians have attempted to impose upon its inquiries. But, when science, passing beyond its own limits, assumes to take the place of theology and sets up its own conception of the order of Nature as a sufficient account of its cause, it is invading a province of thought to which it has no claim, and not unreasonably provokes the hostility of those who ought to be its best friends.

For, while the deep-seated instincts of humanity and the profoundest researches of philosophy alike point to mind as the one and only source of power, it is the prerogative of science to demonstrate the unity of the power which is operating through the limitless extent and variety of the universe, and to trace its continuity through the vast series of ages that have been occupied in its evolution.

Made in United States
Orlando, FL
04 November 2026

90205459R00067